WHAT IS SCIENCE?

Science as an Adaptive Capacity

BY

GEOFFREYSON KHAMALA

Published by:

FOUNDATION
"Thinking for the Universe"

Copyright © 2014 Geoffreyson Khamala

All rights reserved.

ISBN-13: 978-1503241589

ISBN-10: 1503241580

DEDICATION

I dedicate this book to thinkers, visionaries, dreamers, artists, musicians, sculptors, practitioners, tinkerers and lay scientists.

TABLE OF CONTENTS

DEDICATION III
LIST OF ACRONYMS & ABBREVIATIONS V
PUBLICATIONS BY GEOFFREYSON KHAMALA VI

ABSTRACT VII

CHAPTER ONE 1

TRENDS IN THE UNDERSTANDING OF SCIENCE 1

INTRODUCTION 1
SCIENCE AS SUPERNATURAL PHENOMENON, MYTHOLOGY AND TRADITION 1
SCIENCE AS TRUE KNOWLEDGE; WISDOM 7
SCIENCE AS A TYPE OF KNOWLEDGE; KNOWLEDGE OF NATURE . 9
SCIENCE AS A WAY OF PURSUING KNOWLEDGE; THE SCIENTIFIC METHOD 10
EMPHASIZING ON PURPOSE TO DECIPHER THE PROPER MEANING OF SCIENCE 15
CONCLUSION 18

CHAPTER TWO 19

SCIENCE AS AN ADAPTIVE CAPACITY 19

INTRODUCTION 19
FEATURES OF SCIENCE AS AN ADAPTIVE CAPACITY 21
CONCLUSION 79

CHAPTER THREE 80

CONCLUSION 80

BIBLIOGRAPHY 82

LIST OF ACRONYMS & ABBREVIATIONS

BC   -   Common Era

BCE   -   Before Common Era

DNA   -   Deoxyribonucleic Acid

GPS   -   Global Positioning System

IAEA -   International Atomic Energy Agency

NATO -   North Atlantic Treaty Organization

NGOs -   Non-Governmental Organizations

NPT   -   Non-Proliferation Treaty

SKA   -   The Square Kilometre Array

UN   -   United Nations

US   -   United States

NIO   -   The New International Order

PUBLICATIONS BY GEOFFREYSON KHAMALA

1. The Perfect Theory: A Complete Unified Description of the Universe (2014)

2. What is science? Science as an Adaptive Capacity (2014)

3. Is Science Religion? (2014)

4. Wither Globalization Enter Connectedness (2014)

5. The Ultimate Theory: The Perfect Description of the Universe (2015)

6. Tajiriba Spaces: The Solution to Sub-Optimal Outcomes (2015)

7. Zero Unemployment in Kenya: The Utility of Tajiriba Spaces (2015)

8. Reclaiming the Sahara: A Case for Universal Connectedness (2015)

ABSTRACT

Science is a human imperative. Humans have always sought to understand the workings of the universe and our place in it. From early on, pioneer scientists speculated that daily life is governed by a universal plan. They hypothesized that all phenomena (human and natural) conform to a general law. They speculated that nature unfolds in a predictable way. However, in the quest to understand and explain unknown patterns in the physical world, science has been described using opposing lens.

Science was once understood as supernatural phenomenon, mythology and tradition. Science has also been explained as the quest for true knowledge (wisdom). Science is also understood as a form of knowledge, that is, knowledge of the natural world. Today science is understood as an approach of acquiring comprehension or rather the scientific method.

These descriptions are almost certainly not precise.

This write-up seeks to demonstrate that science is actually an adaptive capacity. Science represents the incremental advancements in the way humans and other living lives harnesses nature to facilitate continued endurance. Science is prompted by the trueness that humans and other living lives have never come to terms with their fate.

Science is an essential part of the totality of (human) experience. Since the dawn of life, humans have been pre-occupied with grasping the fullness of the vast unknowns so that they may possibly manage to travel back in time, survey idiosyncratic futures and in effect become immortal.

Science gives us more control over the world around us. Technology facilitates the manipulation of the natural world to perpetuate life. Accordingly, what defines science is the purpose (or goal) not the method or any other decisive factor.

Science is about change not consensus. The task of science is not merely to improve our understanding of the universe but to change it as much as possible in anticipation of achieving never-ending life.

The purpose of science is to grasp and apply the underlying principle(s) of nature to sustain life. The universe was meant to last by means of unending adaptation. Scientifically put: the edict of the universe is to facilitate meaningful life.

Keywords: Science, Adaptive capacity, mourning, immortality, edict of the universe

CHAPTER ONE

TRENDS IN THE UNDERSTANDING OF SCIENCE

INTRODUCTION

Scientists often ponder what science is, how it functions, and whether it exhibits large-scale patterns, trends and meaning. This chapter traces the changing perspectives on the understanding of what science is from early on to the contemporary period.

SCIENCE AS SUPERNATURAL PHENOMENON, MYTHOLOGY AND TRADITION

Since the dawn of humanity the allure of the unknown has been part and parcel of the human society. Pre-historical peoples exhibited scientific curiosity foremost in the form of wonder, mystery, miracle or supernatural phenomenon. The systematic burial of the dead is noticeable a pointer to an almost universal belief that life continue in some form after death and that human beings continue to have needs in the afterlife. People were buried with their courtiers, furnishings and other burial objects such as pottery, tools, weapons, clothing and jewelry, assorted foods and drinks, statues of themselves or others, paintings or models of scenes from daily life that were expected to be useful in the land of dead. The number and sumptuousness of the burial objects reflected the wealth and social standing of the deceased. But even the poorest were accorded preservation rites and provided with foods and fruit to guarantee them survival in the afterlife.

During prehistory (the Stone Age) humans generally lived as nomadic hunter-gatherers. The world's climate became drier and cooler between around 8000 and 6000 BCE ending a period of successful fishing, hunting and food-gathering.

Early humans, during the Neolithic Period, embarked on farming and animal husbandry. The adoption of agriculture, to replace hunting and gathering, as the principal means for obtaining food contributed to the emergence of complex societies including civilizations. Mesopotamia, located in the Near East, corresponding to current-day Iraq, Kuwait, Syria, Turkey and Iran, produced the world's first civilization in the river valley of Tigris and Euphrates.

During this early period, religion and myth pervaded all phases of life, justified traditional rules of morality, and provided people with satisfying explanations for the operations of nature and the mystery of death. Natural objects (the rain, the sun, the moon, the rivers, the mountains, lightening and windstorms) were seen either as gods or as the abode of gods. It was assumed that natural disasters such as droughts, floods, famines, earthquakes, etc. were divine punishment for alleged transgressions.

The political life of the Sumerians, founders of urban life of Mesopotamia, was theocratic. People regarded their rules as either divine or representatives of the god and believed that law originated with gods - divine kingship. Ancient Mesopotamian concept of Justice was Hammurabi's Code of Laws issued around 1750 BCE. The prologue to the legal texts reveal that the Babylonian ruler, Hammurabi (1792-1750 BCE), was chosen by a god to uphold justice in the land, to punish the wicked, and to further the welfare of the people. The laws of Hammurabi and many other pre-Biblical laws (commonly described as an 'eye for an eye' system of justice) defined various types of crimes and the appropriate penalties to be applied.

Near Eastern art and literature were denominated by religious themes. They invented stories (myth) about the gods' birth, deeds, death, and resurrection. For example, *The Epic of Gilgamesh* written about 2000 BCE is about Gilgamesh, a historical figure, who ruled the city of Uruk about 2600 BCE, who engaged in the unparalleled human protest against death. However, towards the end, Gilgamesh learns to accept that death is inescapable (Kovacs, 1989).

Ancient Egypt was also characterized by divine rule and theocratic kingship and the notion of exceptional human beings (agents of the gods). According to Ancient Egyptians, their king or pharaoh was both a god and a mortal. The king was regarded as a benevolent protector who controlled the flood waters of the Nile, kept the irrigation system in working order, maintained justice in the land, and expressed the will of the gods in life and in death.

The Egyptians invented effective methods of embalming to keep intact dead bodies (especially those of their kings) for thousands of years. Dead bodies were buried with objects such as hunting and farming equipment, food items and even their servants for it was believed they will be needed by the deceased.

The ancient Mesopotamians, Egyptians and other ancient civilizations recorded the activities of gods and their human agents in the form of myths, legends, and fables, among others. Nature and human experience was interpreted through mythology and traditions. All occurrences were attributable to the actions of gods, whose behavior was often unpredictable. Bad experiences such as death were seen as the will of gods and demons who manipulated events according to their desires. Mythology and traditions made the universe and life intelligible. Mythical explanations of nature and human experience made living life seem less overwhelming and death less frightening.

The ancient blind poet Homer, who probably lived during the eight century (around 900 BCE), authored two works christened *The Iliad* and *The Odyssey*, describing the events of the Trojan War thereby developing the humanistic outlook.

For Homer, human existence has a pattern. Human affairs and life is governed by a universal plan. People, even the gods, operate within a certain fixed plan. All events (both natural and human) obey a universal law. This idea was later expressed in scientific terms marking the beginning of the quest to understand and explain patterns in the physical world.

In retrospect, Thales in the 6th century BCE rejected supernatural, religious or mythological explanations for natural phenomena proclaiming that every event had a natural cause.

The ancient Greek travel writer Herodotus (484 BC – 425 BCE) who is considered "the father of history" in *The Histories* teaches that thoughtless self-importance (hubris) invites the wrath of the gods. Herodotus chronicled the history of the Persian wars (492–449 BCE). Greco-Persian Wars were a series of wars fought by Greek states and the Persian Empire (now Iran, Pakistan, Afghanistan, Tajikistan and Azerbaijan).

Herodotus hoped to fathom human nature by drawing lessons from the study of the past. He concluded that the political independence of the Greeks ended because of Athenian pride.

On the contrary, Thucydides (460-395 BCE) the originator of scientific history in *History of the Peloponnesian War* did not acknowledge divine intervention in human affairs. For Thucydides, history is the consequence of the choices and actions of human actors.

Thucydides' approach, following Thales and Homer, marks the turning point from mythology and tradition to reason. Thucydides examined activities of human beings and their motives for what they were and unambiguously rejected heavenly explanations for human occurrence. He sought for natural causes and based his conclusions on evidence. He wrote to provide lessons about human behavior in political situations for the benefit of future political leaders and military figures.

The Peloponnesian War (431-404 BCE) was fought between the Delian League (Athenian empire) led by Athens and Peloponnesian league led by the Spartans. The Peloponnesian league was a coalition of Thebes, Corinth and Sparta.

Thales, Homer, Thucydides among others were the forerunners of the Socratic conception of the rational individual.

SCIENCE AS TRUE KNOWLEDGE; WISDOM

Socrates (469-399 BCE) as expressed by Plato in *The Apology* suggested that science is the pursuit of true knowledge - wisdom.

Socrates sought to know more about human nature, the nature of political communities, and human knowledge itself. For him, rational discussion between two or more individuals (The Socratic Method) was the preferred path to wisdom (virtue and self-knowledge). The Socratic dialectic method challenges one to produce reasons for one's view. Wisdom is true knowledge. Philosophy is the love of wisdom.

According to Socrates, "Wonder is the beginning of wisdom". Socrates was dismissive of earlier forms of knowledge for being wholly metaphysical, and lacking in critical reflection.

Plato following in Socrates footsteps defined knowledge as "justified true belief." Confucius agreed with this position thinking thus: "Real knowledge is to know the extent of one's ignorance." For Confucius, wisdom can be learned via reflection, imitation or experience. True knowledge is analogous to humanity. In Taoism, wisdom is closely associated with charity, simplicity, and humility.

Socrates and his predecessors rose above mythology, magic, miracles, mystery, and custom to assert that reason was the sole avenue to true knowledge. They defined human beings by their capacity to use reason thereby developing a rational-scientific outlook devoid of mythical interpretations of nature and the human community. They proposed physical explanations and gradually omitted the gods from their accounts of how nature came to be the way it is. They analyzed government, law, and ethics in logical and systematic ways.

Many Greek city states, particularly Athens, developed democratic institutions and attitudes and came to understand that law was the product of human reason. Plato in *The Republic*, preferred philosopher kings to be leaders of the ideal republic.

Plato in *Phaedo* recounts Socrates' last days moments before drinking the poison. Socrates appeared confident for he was certain he would survive the demise of his body. For him, life is cyclical. Death is simply a phase in life. There is another life after death.

SCIENCE AS A TYPE OF KNOWLEDGE; KNOWLEDGE OF NATURE

Aristotle (384-322 BCE) originated the notion of science as the knowledge of the natural world. For him, reliable knowledge is about the working of natural things.

In his *Metaphysics*, Aristotle saw knowledge as the understanding of causes. Knowledge should be type that can be logically and rationally explained. He valued reason tampered with concrete details of nature obtained through sense experience. He observed that scientific thinking encompasses both rationalism and empiricism. He believed that universal truths could be reached via induction.

Following Aristotle, Nicolaus Copernicus, Johannes Kepler, Galilei Galileo, Francis Bacon, Edmond Halley, René Descartes, Robert Hooke and later Isaac Newton, Gottfried Wilhelm Leibniz, Blaise Pascal, Albert Einstein, Immanuel Kant, Thomas Hobbes, John Locke and Charles Darwin, gave birth to modern[1] science by cementing the idea of science as the pursuit of knowledge of nature.

A philosophical interpretation of nature was gradually replaced by a scientific approach using inductive methodology.

SCIENCE AS A WAY OF PURSUING KNOWLEDGE; THE SCIENTIFIC METHOD

Today the word "science" has become increasingly associated with a way of knowing; the scientific method. The scientific method seeks to explain the events of nature in a reproducible way by means of systematic observation, measurement, experimentation, and the formulation, testing, and modification of hypotheses.

---

[1] Modern period refers to post-medieval Europe

Modern science, therefore, is a deliberate engagement that creates and systematizes knowledge in the form of testable explanations and predictions about the universe and that substantiated by factual evidence. This means that the knowledge must be based on observable phenomena and capable of being tested for its validity by other researchers working under the same conditions.

The Scientific Revolution that took place in 16th and 17th century Europe (early modern period) marks the dawn of modern science. The scientific method is fundamental to modern science such that some consider previous (pre-modern) inquiries into nature to be pre-scientific.

This disciplined way of studying the natural world takes two forms: natural and social sciences. Natural sciences study natural phenomena of the material universe and their laws (including biological life). Natural sciences rely on experimental, quantifiable data and focus on accuracy and objectivity. Social sciences study human behavior and societies. Social sciences are reliant on qualitative research.

Mathematics, statistics, logic and computer science are often classified as formal sciences because the methods of verifying their knowledge are a priori rather than empirical. Also, the connection between formal sciences and reality remains ambiguous. As such, whether these disciplines are properly classified as sciences is contested. Some contributors do not see these disciplines as scientific. However, though formal sciences are not considered "proper" sciences, they provide many tools and frameworks used within the natural and social sciences. For instance, natural and social sciences rely heavily on mathematics as the logical framework for formulation and quantification of principles.

Modern scientists also demarcate natural and supernatural explanations. Supernatural explanations, it is suggested, are matters of personal belief and as such outside the scope of science. To be termed scientific, a discipline (or method of inquiry) must be based on observable and measurable evidence and subject to specific principles of reasoning and experimentation.

Strict interpretation of science as the scientific method means that all products of the human mental processes, including religion, mathematics, statistics, computer science, philosophy, literature, painting, drawing, drama, poetry and art are non-scientific disciplines.

This reality has prompted a non-ending concern and controversy on what scientific knowledge is and whether it exhibits large-scale patterns and trends.

Major positions on scientific progress are associated with Thomas Kuhn, Karl Popper and Paul Feyerabend.

Thomas Kuhn in *The Structure of Scientific Revolutions* (1962) attributed changes in scientific theories to changes in underlying intellectual paradigms ("paradigm shifts"). According to Kuhn, scientific knowledge moves through competing dominant paradigms or structures of thought and practices and is not necessarily progressive. In Kuhn's model, conceptual frameworks are simply different ways to think about the world. New models discard the old way of looking at the universe, and comes up with their own and guidelines for understanding the universe.

On the contrary, Karl Popper argued that scientific knowledge is progressive and cumulative.

Popper proposed falsifiability (falsification of incorrect theories) to replace verifiability. A theoretical framework must be falsifiable to be scientific. He suggested that scientists work on a trial and error basis and adopt theories that are more accurate versions earlier models or that are progressively closer to truth. In Popper's approach scientific progress is a linear accumulation of facts, each one adding to the last. He suggests that all human knowledge is fallible and therefore uncertain.

Then again, Paul Feyerabend in *Against Method* argues that scientific progress is not the result of applying any particular method by all scientists at all times. For Feyerabend, 'anything goes'. For him, scientific knowledge is neither cumulative nor progressive and that there can be no demarcation in terms of method between science and any other form of investigation such as witchcraft.

Feyerabend observed that falsification is never strictly followed noting that often scientists still hold on to theories have failed many sets of tests. Feyerabend vouched to a pluralistic methodology observing that many forms of knowledge which were previously thought to be non-scientific were later accepted as a valid part of scientific canon.

EMPHASIZING ON PURPOSE TO DECIPHER THE PROPER MEANING OF SCIENCE

Science is a process of exploring the familiar and unfamiliar universe around us. The universe appears to be intricate and incredibly complex. Science helps us to appreciate, describe and tame the world.

Science is basically life-forms' capacity to grasp and control the universe. Science is the set of empirical, theoretical, and practical knowledge about the natural world. Technology is the ability by life-forms to manipulate the natural world to perpetuate life.

Science is so uncertain such that people learn from unlikely experiences. This is why curiosity[2] is an indispensable component of the scientific edifice. Curiosity enables to us to comprehend, master and control our surroundings to maintain life.

Purposeful science is collaborative, inclusive and goal-directed. The way scientific ideas, discoveries and breakthroughs are shared is shifting as the Internet becomes ubiquitous, faster, and progressively more accessible.

---

[2] Curiosity refers to inquisitive thinking; desire to know

Scholarly communication is evolving. Whereas all scientific results must be replicable, not everything that can stand up to scrutiny is worth mentioning. Similarly, not everything that is premised on little solid evidence should be dismissed without further scrutiny.

In this age of very exchangeable information, scientists have a responsibility to communicate their work with the wider public. Though peer review has been regarded as crucial to the reputation and reliability of scientific research, it is also widely acknowledged that traditional peer review has severe limitations. Some scholars even doubt peer review's overall efficacy.

Scientific codification evolves, so do tools. Post-publication peer review has now been embraced to ensure more scientific findings are shared publicly in the face of constraints of journal publication ratification by the scientific community.

Science is basically the search for a better appreciation and understanding and the subsequent acquisition of the ability to manipulate nature. Human beings do not just seek to know systematically rather they wish to control their destiny and that of the rest of the universe.

Science has a goal that the scientific activity seeks to fulfill. The method only increases the chances of success. Therefore, science cannot be equated to the method. Science cannot be reduced to merely making testable predictions.

In the research process there normally three pertinent questions: Does the researcher have a problem? – Purpose; is the problem researchable? – Method; and is the researcher capable of carrying out the study? – Training (skills and knowledge) and Resources (time, finances, etc.).

When thinking about undertaking any research, scientists always ask: what mischief will the new findings cure? Unless this criterion is satisfied the research cannot proceed. This is because methods can be varied, training can be enhanced and alternative resources can be found but not so with why the research is being undertaken in the first place.

Science encompasses knowledge accumulation, transmission, application and perfection. Often, the conformist rules of academic tenure and promotion treat citation metrics in specialized journals as the ultimate appraisal for accomplishment. Research influence can also be measured by industry income (innovation), change in human relations and value system, etc. Against this backdrop, citation counts are not a good measure of scholarly success.

The standard expectation is that scientists should generate knowledge that can solve key societal problems and challenges. Standout scholars are expected to reach and engage the community of scholars and the broader public on the problems society faces and advance possible solutions.

CONCLUSION

The history of science has been marked by a chain of advances in knowledge and technological innovations which have inspired a more accurate understanding of what science is. This chapter concludes by pointing out that our evolving perspectives of what science really is compliments our daily struggles in the search for the meaning of life. Against this backdrop, the next chapter further develops this argument to concretize the idea that science is actually an adaptive capacity.

CHAPTER TWO

SCIENCE AS AN ADAPTIVE CAPACITY

INTRODUCTION

This chapter originates the notion that scientific enterprise is essentially an adaptive competence. Science is fueled by inquisitiveness and imagination on the possibility of infinity. It is this curiosity that prompts inquiring minds to contemplate on the ultimate origins of humanity, the size of the extended universe, the geometry of the cosmos, the practicability of interstellar flight, the likelihood of expanding the human presence beyond planet earth, the prospect of a subterranean land inside the earth, the existence of aliens, the end of the world, the prospect of intercepting catastrophic asteroids, the travails of artificial intelligence, the existence of God, the possibility of time travel, the practicability of sustainable habitats and the odds of a perfect theory that will make everything fit together.

Life and non-life are inextricably intertwined. Humans, their technologies and the rest of the universe share a common goal. Science, technology and natural processes are adaptive capacities to sustain life. This explains why safety features are key components of any technology. For example, cars are increasingly coming with built-in sensors that can avoid accidents and even self-park. There are many more examples out there.

Science when understood this way captures every conceivable facet (i.e. social, political, economic, technological, and psychological) of human or any other living life and the processes involving the non-living.

In the holistic sense, therefore, science is what is done in academic institutions, research organizations, professional societies, industry, corporations, governmental institutions, non-governmental institutions, civil society and all other everyday activities in the universe. We are science. When it rain objects it is science as a practical activity, and when we ponder why it rain objects it is science as an academic engagement.

Science seeks to grasp the common thread that unites the universe including everyday practical activities and the attempts to comprehend all there is in order to grow and expand human knowledge to protract life.

FEATURES OF SCIENCE AS AN ADAPTIVE CAPACITY

The major characteristics of science are that:

i. Science is both scholarship and experience;
ii. Science is evolutionary;
iii. Science is random phenomena; and
iv. Science is partial (i.e. goal-oriented).

i. Science is an Academic and Practical Activity

Science in the holistic sense is both an academic pursuit and a practical pastime (Khamala, 2014). The universe (physical world) embraces all human activities/events and natural phenomena (non-human activities/events). The scientific enterprise as practiced in the academy is a human event. Volcanic eruptions, earthquakes, storms, rainfall, lightening, death, and many others, are natural phenomena.

Considering these insights, science as studied in the academy represents only a fragmented view of the world. No wonder Immanuel Kant (1724-1804) in his work *Critique to Pure Reason* argues that the noumenal (true) world (the thing-in-itself) is not knowable instead the subject for scientific study should be the phenomenal world (phenomena). In Kantian terms, sensory and mental representations are mere phenomena.

Science as an academic activity encompasses natural, human (humanities), social, formal and other sciences. Other sciences refer to a plurality of fields outside of the "proper" sciences and include child play, fantasy, science fiction, witchcraft, magic, telepathy, energy healing and astrology, among others.

Knowledge in science is gained by a gradual synthesis of information from diverse sources, by various players including academic researchers and across different domains of science. Tinkers, artists, sculptors, painters, musicians and other performing artists, and lay people are vital participants in the science project.

Knowledge comes from experience. Humans (in their capacity as individuals and as the sum total of all humans) do not possess a monopoly over knowledge. The universe itself is a knowledge system. That is why biological systems may possess knowledge without being conscious of it.

Experience is the source of knowledge. Not all knowledge is empirical (sensation) however. Therefore, science cannot be tied to empirical verification.

The scientific enterprise acknowledges five domains of knowledge: (i) the human brain (individual beings); (ii) named human society (human culture); and finally unconscious natural processes, that is, (iii) the immune system; (iv) the DNA genetic code; and (v) the physical world (the universe as an intelligent system).

This is an improvement to Karl Popper's supposition of three worlds of suitable epistemological spheres of knowledge: the objective sphere (world one), the subjective sphere (world two), and the inter-subjective sphere (world three). Contra to Popper, science as an adaptive competence privileges connectedness.

Human beings are part of the intelligent universe. The science project enables us to work together to create, share and apply new ideas. Knowledge is not an end but rather a means to an end. Scientific method, the body of techniques for investigating phenomena, acquiring new knowledge, or correcting and integrating previous knowledge is an important ingredient in scientific inquiry but it is not equivalent to science.

The scientific method is an adaptive feature of science to ensure any research activity measures what it purports to measure (validity) and that the results are consistent (reliability). Other adaptive features include possession of requisite skills (experience), relevant academic qualifications, journals and peer review mechanism. What makes an activity scientific is the goal not the method. The scientific method only ensures that we have well-supported findings.

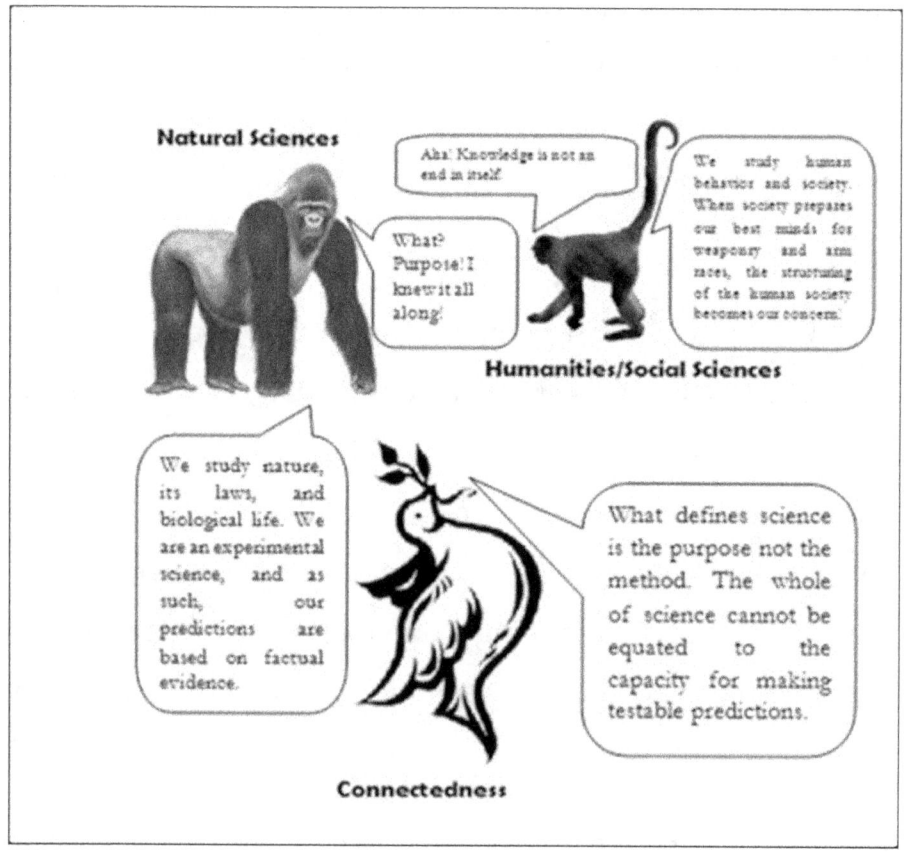

Fig. 1: Science as a Whole

The goal of science is to rewrite the boundaries between life and death. Science is in essence the acquirement, improvement and employment of (human) adaptive capacities to safeguard life and circumvent death. The emphasis of the scientific enterprise is on the connectedness between humanity and the rest of the universe.

Connectedness is what impels science. Science is simply the extension of the physical world. We are science. The subject matter of the totality of science is all encompassing: natural phenomena; and human activities, behavior and mental processes.

Human activities encompass situation-specific human behavior; the functioning of social groups; mental faculties that include religion, language(s), political systems, economic systems, monetary systems (currencies), theoretical and analytical constructs, imaginative works such as stories, poems, and plays, performing arts such as acrobatics, comedy, dance, magic, music, opera, film, and theatre; among others.

ii. Science is Evolutionary

Non-life evolved into life. Since life appeared on Earth, living life continually adapt to their current surroundings to avoid a relapse into non-life and more. All organisms adapt to survive in their ecological niche or habitat.

Adaptation operates through natural and sexual selection mechanisms (plus genetic drift, gene flow and mutation). Adaptation also occurs through human intervention (mental activity, application and perfection of technologies). Adaptations may be structural (morphological – how the organism is built), functional (physiological – how an organism works inside) or psychosomatic (behavioral – what an organism does).

Morphological adaptations include the size, shape, colour and structure of an organism. Physiological adaptations are systemic responses of an organism in response an environmental stimulus resulting in the improved ability of that organism to cope with its changing environment. Organism perceives the world through physiological capacities such as senses. Non-human organisms may possess senses that are absent in humans, and are able to sense the world in a way that humans cannot. Subsequently, some living organisms other than human beings have evolved a multitude of coping (survival) strategies to survive extreme environments by being altitude, cold, chemical, dry, fire or heat tolerant.

Survival strategies enable organisms to cope with environmental changes and the constant threat of predation.

Human beings are endowed with psychosomatic adaptations mental faculties and emotions. Psychosomatic adaptations can be inherited or learnt. Mental faculties make human beings unusually adept at learning new skills or using tools, language development, and socio-political institutions, among others. Humans are also endowed with highly developed social and emotional skills. Emotions can be observed in empathy, compassion, mourning, cooperative aspects, altruism, aggression, courtship display and parenting.

An organism's morphological, physiological, behavioral and other properties are determined by its genes and its environment. Individual organisms retain the same genes throughout their life. However, populations evolve over time such that a single organism is never typical of an entire population. The distinguishable differences in a population are due to modifications in successive generations in the hereditary gene pool as the population adapt to their current surroundings.

In the struggle for life, organisms transform from simple life forms to more complex life forms through natural processes. The fossil record supports this supposition. Evidence also abounds in the form of human adaptations such as anatomic changes for language capacity. However, organisms are not passive targets of their environment. Living organisms shape, sustain or maintain the habitat around them. Species do not simply change to fit their environment; they modify their environment to suit them as well.

Behaviors have evolved over time, similar to the way that physical traits are thought to have evolved. Human adaptive capacities such as the incest taboo, socio-political institutions such as the family, lineages, clans, tribes, states, the state system, the invention of money, and the development of formal religion seem to suggest that life and death structures the human society. These adaptive capacities have proven to be evolutionarily successful over time, that is, they are conducive to the perpetuation of life.

However, it is evident that human beings (and other living life) are an ambitious lot. Living life hopes to extend their lifespan for an indefinite period. So far, only Planarian worms have the capacity to regenerate indefinitely. This has raised the possibility of alleviating ageing and age-related characteristics in human cells.

(a). Political evolution

Markers of identity are apparent in the structuring of the human society (Khamala, 2009). Gender is the most decisive marker of identity for many people around the world. Gender segregates individuals, groups and the world society at large. Without doubt, gender informs the construction, reinforcement and persistence of the family, kinship/lineage, clan, ethnicity, state, race, civilization, religion, status and caste. Age is the only marker of identity that is not directly influenced by gender. Anatomical, functional and behavioral adaptations distinguish male and female sexes.

Gendered power units are constructed and reproduced because of the way society is structured (that is, the unequal relationships between men and women) right from the family level.

The family is one of the most captivating creations by humankind. The family is meant to provide emotional ties, the start of life (mating system), and an insurance for every member of society to find someone to mate with. However, the most evident aspect of the family is the restrictions on who can have sex with whom (the incest taboo) and the consequent identification of family members and relatives like father, mother, cousin, aunt, uncle and so on.

The incest taboo is almost universal. The incest taboo is an adaptive feature to preempt detrimental consequences of inbreeding; a deliberate human attempt to forge valuable alliances among otherwise unrelated households; and a way to avoid conflicts arising from competition among close (and sometimes distant) relatives over mating partners.

The survival of humanity rests on the restriction of sex among its members and the consequent need for other family units as sources of spouses. So, preserve life, family members are forced to reach out for non-relatives. As such, present-day (blemished) civilization would not have happened without the family.

The family identity is attained through the father or the husband particularly the inheritance of surnames. Women do not bestow any form of identity. This prompted Michel Foucault (1926 - 1984) in *History of Sexuality* (1978) to problematize sex and power relations within the human family. He noted that sex is a form of control that helps to define or enhance certain power relations of domination and oppression. The family acts as the building block of lineages (and kin relations).

Reckoning kinship is universal. Kinship reckoning respects the incest taboo. Kinship terminologies distinguish between sexes, generations and blood relatives. Kinship strengthens the bond between the related persons relative to strangers. Rules of kinship and descent have important public aspects, especially under monarchies, where they determine the order of succession, the heir apparent and the heir presumptive. A number of related lineages constitute a clan.

Most clans are patrilineal and patrilocal. Clans are the primary force of security as the clansmen are obliged by honour to avenge one another. The clan remains in use by the royalty and nobility to demarcate hereditary line and dynasty. However, critics often demonstrate that paternity in patriarchal societies is a matter of conjecture. Clans are the building blocks of tribes.

Tribes privilege primordial social ties (Fried, 1972). Tribal divisions shape interactions with others: altruism (positive interactions with unrelated members), kin-selectivity (positive interactions with related members), and violence (negative interactions with non-members).

Without a DNA analysis, many members of patriarchal societies proudly defend their tribes on the basis of their presumptive paternity. Warfare is a typical feature among tribes given the tendency to exploit tribal instincts.

Tribe is the most powerful tool for mobilizing political opinion in many places in the world. Indeed much of the political tension in modern societies is the conflict between the desire to organize a nation-state based on tribal ties. Gullible citizens are exploited by politicians who use tribal affiliation to get advantages in political and economic deals, advantages that they do not later share with their presumed tribesmates.

The distinction between tribes and ethnic groups is to some extent academic. Members of an ethnic group are usually conscious of their membership to the group and others recognize the group's exclusivity. The debate whether ethnicity is universal is unsettled. However, almost every major region of the world has societies separated along ethnic lines. Intra and inter-ethnic relations involve cooperation, inter-dependence and every so often conflict in its diverse forms.

For the most part of human history, people have lived in primordial state-forms until the Neolithic period. The first known states were created in Ancient Egypt, Mesopotamia, India, China, and Ancient Greece. The Ancient Greeks were pioneers in articulating a rational ground for the reality of political institutions. Prior to this, political institutions were described and justified in terms of religious mythology and tradition. Still, the Greek idea corresponds more accurately to the modern concept of the nation—i.e., a population of a fixed area that shares a common language, culture, and history.

The modern state as we conceive of it today is based on a pattern that emerged in Europe in the period from 1100 to 1600. Subsequent to the Montevideo Convention on Rights and Duties of states that signed at Montevideo, Uruguay, on December 26th, 1933, for an entity to qualify as a state it should occupy a definite territory, possess a permanent population, submit to an effective political authority over the extent of such territory and the claimant to statehood must have the capacity to enter into relations with the other states including the capability to fulfill treaty obligations.

Evidently, human beings have experimented with bands, tribes, chiefdoms and state societies (Service, 1975). Today states have almost completely displaced primordial ways of organizing societies all over the planet. Still, the present state system is imperfect because most of the contemporary states are divided and fragmented along fault lines (i.e. family, lineage, clan, ethnic, racial, religious, and civilizational persuasions).

Informal networks such as familial and ethnic obligations constrain the capacity of states to fulfill their adaptive functions. The influence of primitive state forms run contra to the 20th century idea of citizenship (in a sovereign state). In fact, notions of sovereignty originate from the human insecurity over the real intentions of other primitive human groups christened states.

Right from its early origins, the state is still a contested arena. Socrates (469-399 BC) thought that it was possible to reform the individual by cultivating the employment of the human faculty of reason. On the contrary, Plato (423-347 BC) noted that it was hard for the individual to undergo moral transformation while living in a depraved and crooked society. He proposed reforming the state thereby preempting the present-day constitutional and legislative frameworks (positive or man-made law). Constitutions and legislation are adaptive capacities to tame primordial instincts. The police have evolved to deal with internal security while the defense forces deal with external insecurity. States constrain individual choices and pool the synergy of its members to pursue common goals.

State arming including acquisition of nuclear weapons is an adaptive capacity. The acquisition and deployment of weapons by states is informed by the human fear of demise and the subsequent quest for self-preservation. States always arm themselves to secure their domestic and foreign interests in an environment of anarchy (in the absence of a worldwide government). Yet arms acquisition by a state to increase its security decreases the security of other states and the ensuing security dilemma (Herz, 1951). Such a scenario may intensify tension and contribute to an arms race or even war.

The human mindset imagines the enemy to be the citizen or another state necessitating the current police and military apparatus. Historically, human insecurity has prompted states to enter into precarious alliances (often informed by primordial instincts and/or vested interests) and global approach that have fuelled rather than quelled insecurity.

The beginning of warfare is traced to the prehistoric period. Nonetheless, warfare as an adaptive capacity in the nuclear age is no longer a viable option. Already 26 states in the world including Haiti, Grenada, Costa Rica, Samoa, Kiribati and Vatican City do not maintain armed forces.

The formation of the United Nations (UN) was an adaptive capacity to save future generations from war, reaffirm human rights, establish equal rights for all persons, and promote justice, freedom, and social progress worldwide. The UN was formed by Allies (Great Britain, the United States, and the Union of Soviet Socialist Republics) with the support of other countries during the Second World War.

The sticking point at the UN is the UN Security Council, the most powerful of all the branches, composed of five permanent members and ten rotating members and mandated to deploy UN member states' militaries.

Primordial ties and insecurities were decisive in the composition of the UN Security Council. These primordial alignments and fears precipitated the Cold War (September 2, 1945 to December 25, 1991) between the United States and its NATO allies (loosely described as the West) versus the former Soviet Union and its Warsaw Pact allies (loosely described as the Eastern Bloc). An open confrontation never actually happened probably due to the fear of mutual annihilation as the two group possessed nuclear capabilities. The conflict was fought for the most part ideological.

The aftermath of the Cold War included the disintegration of the Soviet Union in 1991 and the subsequent collapse of communism.

Warfare will continue to a feature of humankind until a global citizenry pursues a common global defense strategy (collective security) that overrides state boundaries and other primordial instincts.

Race is another adaptive capacity. Race is traced to European conquests beginning in the 15th century. Race was used for self-aggregation and to overcome the perils of nationalism (or ethnicity) that had turned Europe into a blood bath. Many historians condemned nationalism as the great vice of Europe. So, Europe used the concept race to externalize its conflicts. Beliefs about race (racism) were used to justify centuries of European discrimination, slavery, imperialism and colonialism of the Americas, Africa, Australia, Oceania and Asia.

Racial essentialism lost scientific credibility when during the 1930s and 1940s, Adolf Hitler propagated the idea he was a member of an Aryan master race that was destined to rule the world through armed conquest. However, the racist ideology still prevails in the form of self-segregation. People self-segregate to avoid integrating with the rest of the population, either to protect perceived advantages or to play victim.

The apartheid system in South Africa was an extreme example of the practice of marrying within a group (endogamy) to encourage group affiliation, bonding and to resist integration with surrounding populations.

The notion of civilizations is another adaptive capacity based on ethnicity (especially primordial kinship ties christened culture/religion) to engage in a zero-sum competition for limited resources, influence or to project power.

The most obvious civilizational power games currently involve the "West". The "West" is an amorphous term to represent the United States and its NATO allies during and after the Cold War and deployed to create the impression that the current global power matrix (with a population of people of European descent or of Anglo-Saxon heritage at the apex) is merely a continuation from antiquity. The notion that the

"West" stretches from Greco-Roman civilization in Europe (including the advent of Christianity) to today's globalization has helped tone down conflicts in Europe, North America, Australia and other parts of the world. But in the face of global terrorism, brain circulation and connectedness, civilizations, with its racist undertones, as an adaptive capacity has become more of a liability than an asset necessitating a rethink.

Religion is the foremost adaptive capacity right from the prehistoric times. The focal point of most religions is death and the possibility of immortality. Religious disputes have been at the root of many historic conflicts – the jihads, crusades, inquisition and terrorism. History is replete with instances of unkindness and injustices perpetrated in the name of religion because organized religion goes hand in hand with a problematical kinship system. In fact, religion is a sufficient reason to list a group as a separate ethnic group. For example, the world's major religions trace their ancestry to the patriarch Abraham. It is, therefore, not surprising that Abrahamic religions are associated with a violent genocidal heritage motivated by an exclusivist ideology that predictably fosters violence against those that are considered non-members.

Political institutions (as informed by unequal gender relations) are constructed in such a way that they ensure the survival of members of the group (while excluding outsiders). Gender informed identities only attempt to project competition and conflict on the outside but fails to eradicate such competition and conflict entirely. Therefore, despite the externalization of enmity, people compete, conflict and die within and between the various constructed socio-political identities. Several movements may change all this: the gender movement, issue-based politics (perfect state/democracy), the need to fix economic fluctuations (perfect economy), the rise of civil society organizations, and the rise of a global citizenry (the perfect state).

The evolution of political institutions is part and parcel of an intellectual thought process to sustain life. Through the science project humans beings are an extension of the intelligent universe. The universe was built to last forever.

Politics as informed by family, kin/clan, ethnic, linguistic, religious and cultural divisions often favour the male gender. Nevertheless, kin/marriage based arrangements are slowly but surely giving way to complex human organizations based on merit and achievement worldwide. Issued-based politics criminalizes force, fraud, pretence and overtones informed by primordial tribal instincts. Subsequently, we are witnessing the evolution of connectedness (human contact worldwide).

Without doubt, a global citizenry is beckoning. The entire world is shifting away from independent, disconnected sovereign states to a world of increased universal awareness and connectedness. The rise of a global citizenry (collective security and perfect democracy) may invoke greater cooperation among human actors for the benefit of all as opposed to national chauvinism, ultranationalism, jingoism, state aggression, religious exclusiveness and civilizational power games.

(b). Economic evolution

Over time, humans have enhanced their ability to harness nature for continued existence. The evolution in the production, distribution, and consumption of wealth began with the discovery of markets. The occasional and spontaneous exchange of goods and services (barter trade system) was replaced over time by deliberate trade structures. This process was accelerated by the introduction of money (a pricing mechanism).

Besides markets and a pricing mechanism, other artificial adaptive capacities that have contributed to economic evolution include technology; war; marriage, and associated primordial ties and sentiments; states and consequent notions of comparative advantage, positive balance of trade; and scarcity.

The natural adaptive capacities that have contributed to the economic evolution include natural calamities such as floods, drought, earthquakes; continental drifts and tides; rainfall; desertification; and climate change, among others. Governments and private business or individual enterprises have played mixed roles in the process.

Early humans practiced hunting and foraging for survival. Throughout the Paleolithic (Old Stone Age) period people generally lived as nomadic hunter-gatherers. Foraging tribal bands were small, relatively non-hierarchical and mostly self-sufficient communities. As there was no surplus product, food-sharing was a safeguard against the failure of any individual's daily foraging. Non-monetary exchange involved the innovative barter, gift, and debt. This type of adaptive capacity was unsatisfactory because often early humans were at the mercy of nature, relied on others for their survival in times of frequent hardships and their awareness of connectedness was slight.

The Mesolithic (Middle Stone Age) period was a transition from gathering and hunting for food to farming and animal husbandry. Nomadic pastoralism was the predominant form of social organization during this period after the world's climate became cold and dry ending a period of successful hunting-gatherer lifestyle. It was during this period that the current familiar global pattern characterized by deserts punctuated by fertile areas emerged.

Early humans adapted largely by establishing settlements around dwindling resources such as rain pools, green valleys and rivers in dry regions. The result was simple farming communities. Forest crop growing (simple farming) was the earliest form of agriculture.

The Neolithic (New Stone Age) period which occurred about the 10th millennium BC, ushered in a sedentist lifestyle with early humans adopting rudimentary domestication of plants and animals. The Sumerians began farming around 9500 BC contributing to the development of early villages. Simple farming and animal husbandry made possible for the emergence of complex farming societies, labour specialization, trade, social classes, war amongst adjacent cultures, and the need for collective action for building dikes, reservoirs, irrigation systems and complex infrastructures such as pyramids triggering the progression from early Neolithic villages to the first cities and civilizations.

Some of the earliest known civilizations arose near river valleys, such as the Euphrates and Tigris valleys in Mesopotamia (later Iraq) — 5200 BC, the Nile valley in Egypt — 3000 BC, the Indus valley in the Indian subcontinent — 3000 BC, and the Yangtze and Yellow River valleys in China - 2000 BC.

Slavery is an age old institution in the world. The practice can be traced back to the earliest records such as the Code of Hammurabi (1760 BC) when it was an established institution. Slavery was an adaptive capacity emerged that when people belonging to the higher classes started treating their fellow human beings as a form property to be directly possessed. The reasons for the rise of slavery include the need for labor for agriculture, commerce, personal and political reasons.

Slavery was not important when humankind depended on food gathering, hunting and fishing as slavery depends on a system of social stratification. Nevertheless, the introduction of agriculture led to the emergence of increasingly more hierarchical social structures. Established traders employed slaves as porters, merchants and trading agents. Besides, early state forms extracted economic surplus produced by local communities in the form of tributes and collective work to finance public infrastructure, especially irrigation systems. Slaves were also needed for military and administration purposes; to perform domestic chores; for sacrifice; and for prestige, power and procreation.

Slavery was not voluntary. Slaves were acquired internally or externally through slave trade (market supply), warfare, raiding and kidnapping, punishment, tribute and pawning.

Sooner or later, social classes became pronounced especially with the rise of kingdoms and empire building. For example, in Ancient Egypt there existed rigid class structure with the pharaoh at the top followed by priests, artisans, farmers and slaves. The higher classes acquired their wealth through the extensive use of animals in agriculture, advanced trade networks and public offices. In some parts of the world slavery evolved into feudalism.

Feudalism was an adaptive capacity that flourished in medieval Europe between the 5th and 15th centuries.

Medieval society was structured around relationships derived from the holding of land in exchange for service or labour.

The Feudal System was introduced to England following the invasion and conquest of the country by William I (The Conqueror). The system had been used in France by the Normans from about 900 AD. The King owned all the land, reserved one quarter to himself as his personal property, some was given to the church and the rest was leased out to trusted powerful and wealthy men (Barons).

The Barons had to pay rent and provide the King with Knights for military service whenever need arose. Knights were given land by a Baron in return for military service. Knights also protected the Baron and his family from invasion. The Knights distributed some of their land to serfs who in turn provided free labour, food and service. Serfs had no rights and were poor.

The market economy traces its origin in Europe in the 16th century, and later extended to most of the world. The market economy as an adaptive capacity has passed through numerous transitions: agrarian capitalism (market feudalism), mercantilism, industrial capitalism, monopoly capitalism, colonialism, welfare capitalism, mass production and mass consumption, state capitalism, corporatism and financial capitalism. In our day, the financial parts of the economy shape the economy. The market economy is an adaptive capacity to facilitate contact, exchange and value allocation.

The alternative to the market economy is the planned economy. In a planned economy, property ownership and wealth distribution are subject to control by a collective for the benefit of the society as a whole.

Supporters of planned economies argue that the market economy is exploitative, promotes inequality and threatens the very cooperative nature of humanity. Critics, however, suggest that planned economies are inefficient in resource distribution. They rightly observe that in planned economies it is difficult to detect consumer preferences leading to periodic situations of shortages and or surpluses; the lack of a price mechanism inhibits informed decision-making, innovation and self-management; and there exist the tendency to suppress democracy, civil liberties and human rights (Popper, 1936; 45; Hayek, 1944).

The principal drawback of the planned economy is the lack of an inbuilt incentive mechanism requiring a collective to plan, manage, control and allocate value. As such, human actors are inhibited to a large extent from utilizing their endurance capacities to harness their environment.

To improve on adaptability, most contemporary economies are mixed in nature whereby both the state and private sector direct the economy. Mixed economies encourage private ownership, are based on the profit motive, have a strong regulatory oversight and have a variety of governmental enterprises and initiatives that promote social welfare. Governments all over the world, exercises sizeable indirect influence over the economy through fiscal and monetary policies designed partly to counteract economic downturns, financial crises and unemployment.

Humans engage in economic activities to exploit their environment for their survival. Humans have been improving on their adaptive capacities (including economic theorizing, innovation, and entrepreneurship, among others). However, the way the society is structured (familial, ethnic, religious and other primordial fault lines) tend to undermine human adaptive efforts.

Nepotism, clannism, tribalism, ethnocentrism, patriotism, racism, colonialism, imperialism, religious bigotry and civilizational jingoism distort human contact, the exchange of goods and services and value allocation. This is why despite scientific progress and in the face of "scarce means having alternative uses" humans still cannot plan on a word scale (Robbins, 1932). It is also the reason as to why our economic reasoning is still parochial.

Our insular thinking lies to us that we are competing for insufficient resources, opportunities and mating partners. Consequently, human social situations and interactions dwell on competition in its diverse forms (i.e. slavery, colonialism, imperialism, conflicts and discriminatory practices such as nepotism, tribalism, racism and apartheid).

Historically, we have tended to dwell so much on the exchange (i.e. production, distribution and consumption) of goods and services and the arising value allocation while ignoring connectedness (human contact and interaction with the environment).

The most important resource of any society is labour (human resource). Labour engages in economic activity to produce wealth. Labour (human effort) is the source of all wealth. Capital is accumulated labour. The world economy is characterized by an excess of labour and shortage of cash (the pricing mechanism). How this turns into the competition for the exploitation of scarce resources instead of resource abundance is a dilemma necessitating a rethink. Money supply is often used to limit production. Scarcity can only be purged in a condition of perfect employment.

Human contact collapses when neglected thereby breeding identity politics. Identity politics poisons markets such that the logic of supply and demand alone does not ensure efficient allocation of resources. No wonder the production, distribution, and consumption of value is often informed by the uncorroborated fear that nonrenewable natural resources such as fossil fuels (i.e. petroleum and coal) and metal ores (.i.e. gold, iron, copper, silver, etc.) are in danger of depletion. This economic reasoning informs all forms of competition (zero-sum in nature) among human groups.

Perhaps this explains why market economies are volatile and always on the verge of collapse due to economic fluctuations and the persistence of suboptimal outcomes such as unemployment. Regrettably, even the best minds draw a blank on this matter.

Perhaps it is because we are constrained by primordial instincts. We dare not imagine a perfect economy that is characterized connectedness: labour mobility/brain circulation – perfect employment; a single world currency; free trade; and global connections, networks, partnerships and integration.

The world deserves more sustainable economies and models – the perfect economy. In a perfect economy there is a possibility of perfect exchange, the best possible allocation of values and the most advantageous human contact and interaction with the environment.

(c). Social evolution

Social behaviour is the way an organism interacts with members of its own species. Human social behaviors range from pack hunting, territorial fights, slavery, colonialism and imperialism, courtship, mating patterns, schooling, parties, music festivals, sports events, interactions on social media, and war, among many others. Social behaviours evolve overtime and determine or depend on the structuring of the society.

Elman Rogers Service (1915 – 1996) in *Origins of the State and Civilization: The Process of Cultural Evolution* (1975) developed a four-stage model of societal evolution implying that all societies started with primitive state forms based on family and lineage structures (hunter-gatherers and tribal bands) before evolving into chiefdoms led by political leaders who inherited their office and then state societies.

For the most part of human history, social interactions have oscillated between cooperation and competition. Human beings have traded, married and interacted peacefully with one another but they have also invaded, conquered, colonized, forcefully converted and warred with one another.

Generally, patriarchal arrangements encourage hierarchy and domination rather than sharing and nurturing. However, today the human society is discarding social behaviours that are incompatible with a living universe especially, cannibalism, infanticide, killing of twins, human sacrifice, dueling, slavery, colonialism, collective punishment, sterilization, geronticide and as anticipated war.

The transformation of social systems, structures and mindsets over time has been accompanied by the development of diverse (re-)enforcement mechanisms such as communal action, change of tradition, emergence of a value system, the application of common sense, sanctions, state legislation, international treaties restricting a state's options in war fighting (including those on disarmament, nuclear non-proliferation, etc.) and establishment of oversight agencies such as the UN and its affiliate agencies such as the International Atomic Energy Agency (IAEA), among others.

Exceptional individuals such as Jesus of Nazareth, Moses, the Buddha, Mohammed, Mahatma Gandhi, Nelson Mandela and Wangari Maathai fought injustice by embracing the human ideal. They emphasized on discipleship and servantship to others, taught patience and tolerance and preached against all types of violence in the society.

The pillars of Islam are the pilgrimage to Mecca (an obligation that must be carried out by every able-bodied Muslim who can afford to and is done at least once in a person's lifetime), faith (*Shahada*), prayers (five times a day), fasting (holy month of Ramadan) and giving charity (amounting to one-fifth of one's wealth). Mecca is the birthplace of Islam whose origin dates back to the time of Prophet Abraham. A Muslim has to redress all wrongs, pay all debts and have enough money to fund the journey and support the family while away, before setting out for the pilgrimage.

(d). Legal evolution

Saudi Arabia may end executions by beheading because there's a shortage of swordsmen in the country.

Oil-rich kingdom mulls abolition of beheading in favour of firing squads for capital punishments due to reported shortages of government swordsmen

Rape, murder, apostasy, armed robbery and drug trafficking are all punishable by death under Saudi Arabia's strict version of Sharia, or Islamic Law.

The beheading issue has always been a source of tension between Saudi Arabia and the international community.

(e). Technological evolution

Technology is the functional application of knowledge. Technology is an activity that requires the practical application of mental and physical effort to produce some value.

Since the dawn of humankind, humans adapt to their natural environments by making, modifying, using and acquiring knowledge of tools, machines, utensils, weapons, instruments, housing, clothing, techniques, crafts, systems, methods of organization, communicating and transporting devices.

The use of tools by early humans was partly a process of discovery and continues with the invention of more complex technologies. This process began with the conversion of natural resources into simple tools.

The three-age system periodization of human prehistory is named for their respective predominant tool-making technologies: the Stone Age, Bronze Age, and Iron Age.

The Stone Age is divided into Paleolithic (Old Stone Age), Mesolithic (Middle Stone Age) and Neolithic (New Stone Age).

The first use of stone tools happened during the Paleolithic period. Other technological advances made during the Paleolithic era were clothing and shelter. Clothing, adapted from the fur and hides of hunted animals, helped humanity expand into colder regions.

New technologies for growing and storing food and for channeling water enhanced people's ability to harness nature.

The discovery of fire marked a major transition in the technological evolution of humankind. By 7000 BC, stone was supplanted by bronze and iron in implements of agriculture and warfare. Metal objects replaced prior ones of stone. Fire broadened the range of foods that could be eaten, its digestibility and its nutrient value. The advancements in technology allowed a more steady supply of food and consumer goods.

The invention of the wheel at around 4000 B.C. revolutionized diverse human activities including agriculture, transportation, production of pottery, war and the application of nonhuman power sources.

Technology is responsible for today's diversified and advanced world economy. Communication was also greatly improved with the invention of the printing press, telegraph, telephone, radio and television. Technological advancement revolutionized transportation allowing powered flight such as the steam-powered ship, train, airplane, and automobile. Advancements in technology led to the construction of skyscrapers and large cities. Technology has made it possible to harness electricity.

Recent technological developments (the internet, mobile phone networks and technologies) have lessened physical barriers to human interaction and communication and allowed humans to interact freely and almost instantaneous on a global scale.

Humans have also been able to explore space with satellites and in manned missions going all the way to the moon. So far six countries namely US, Russia, Europe, Japan and China and India have launched or intend to launch missions to Mars. Curiosity, the most recent orbiter, is on a mission to determine the suitability of Mars for harboring life of its own. Humans have been curious to know whether other planets may once have harbored the necessary habitable environments for microbial life to evolve.

Space missions often unleash immense technological spin-offs. The Square Kilometre Array (SKA) will enable scientists to explore the origins of galaxies, stars and planets. New technology such as genetic engineering, nanotechnology, synthetic biology and robotics have revolutionalized the treatment of diseases and extended the average lifespan.

The acquisition, modification and use of technology are not unique to human beings. Other primates and certain dolphin communities have developed simple tools and learned to pass their knowledge to other generations. Wild chimpanzees utilize tools for foraging. Monkeys also use stone hammers and anvils for cracking nuts. The adult gorilla uses a branch as a walking stick to gauge the water's depth. Bees design and built honeycombs. Parrots are known to use tools in the wild and also possess the ability to craft tools for reaching food and other objects that are out of reach. However, crafting complex organizations such as governments, the military, health and social welfare institutions, supranational corporations, NGOs and civil society organizations is restricted to humans.

Human power over the physical world through technological evolution has its own perils. Humanity has been tinkering with technology that endangers its future existence. Pollution (air, water, and land) is the unwanted byproduct of technology. In nature, organisms reuse the wastes of other organisms. No such mechanism exists for the removal of technological wastes such as toxic waste, radioactive waste and electronic waste. Biggest ecological disasters, such as the Aral Sea, Chernobyl, and Lake Karachay led to the realizations human curiosity may endanger their existence. Long-term harmful effects of technological advancements include global warming, deforestation, natural habitat destruction and coastal wetland loss. The discovery of nuclear fission has led to both nuclear weapons and nuclear power.

Humanity's understanding of the natural world (science) and the ability to manipulate it (technology) have changed over the centuries. The Knowledge and technological know-how increase our adaptive capacity. Considerable scientific effort contributed to the creation of the internet, which has changed the way we interact with each other and with our world. The internet is a place where we can access and exchange our experiences, thoughts and theories about the world. We are now always in a mutual interaction with people, events and knowledge everywhere and anytime.

Conceivably humanity cannot resist the temptation of expanding our knowledge and our technological abilities. Indeed, it appears that technological evolution is essentially beyond the control of individuals or society. Yet technology has frequently been driven by the military. Undeniably, many modern applications developed for the military before they were adapted for civilian use. Humans have since progressed or retrogressed from making and deploying crude weapons such as clubs, machetes and spears to nuclear weapons. Can humanity survive self-annihilation? Without a value system, political, economic and military competitiveness can end in disaster. True, the science project is evidence enough that technological evolution has all along been driven by the urge to survive self and/or collective destruction.

Living things, non-living things and non-things are interdependent. Humans, their technologies and the surrounding environment share a common value system. This is because scientific activity, technological advances and natural events are adaptive capacities to continue existence. This explains may explain away the fear by scientists such as Stephen Hawking and Max Tegmark over the unusual threat posed by Artificial Intelligence (AI). AI refers to the possibility of computer systems being able to mimic human intelligent behavior. The fear is that if and when machines manage to have the capacity to take action on their own without human intervention life as we have to come know it will be destroyed. Actually, the continuation of life in the universe is not necessarily because human beings are in control. Rather our existence is because life is a property of the universe.

(f). Language acquisition and development

Language encompasses the humanity aptitude to learn and deploy complex systems of communication. Humans communicate and share awareness through spoken, written and sign languages. Humans also artificially codify their experience through computer programming. Language is universal and has a biological basis.

Prominent linguists and other scientists such as Pythagoras (570 – 495 BC), Aristotle, Jacques Lacan (1901 – 1981), B.F. Skinner (1904 – 1990), Jean Piaget (1896 – 1980) and Naom Chomsky (1928-), among others have theorized on the acquisition, development and deployment of language.

Human adaptations during the Paleolithic period include anatomic changes bringing to fruition language capacity. The human ability to precisely communicate abstract, learned information allowed humans to share their lived experience thereby enhancing scientific understanding and tool use.

Languages evolve over time. Linguistic change is a universal process. Languages live, die, move from place to place, change and diversify with time. In the distant past, physical barriers such as oceans, high mountains, and wide rivers constituted impediments to human contact. However, present-day technology in the fields of travel and communications has made such geographical factors of less and less account. Political restrictions on the movement of people and of ideas have also been rendered unnecessary.

Humans learn languages, produce and understand utterances because they want to communicate their experiences. Humans defy all odds to communicate their experience foremost being the universal apprehension over bereavement. This explains the origin of language capacity and the emergence of Creole, pidgin, Kiswahili, sheng', etc.

The trend since time immemorial has been the evolution towards a universal language to bring to fruition universe-wide connectedness.

Language diversity distorts connectedness. Having many languages distracts us from communicating simply, clearly and unambiguously our common anxiety over death. Subsequently, we are witnessing the evolution towards linguistic homogeneity to bring to fruition universe-wide connectedness. The world is witnessing prolonged and regular contact between speakers of one or more languages. Possibly a perfect language (a universal language) may emerge to facilitate international intercourse and cooperation, media, commerce, legislation, instruction etc.

The perfect language may not necessarily be today's *lingua francas* such as English, Arabic, Chinese, Russian, French,

Spanish, Swahili etc. but rather a mix from the more than 5000 languages on the planet.

(g). Biological Evolution

Biological evolution suggests that all life is connected and can be traced back to one common ancestor. Life on Earth originated and then evolved from a universal common ancestor approximately 3.7 billion years ago. Early humans (primates) evolved from a species of bipedal foraging hominids (later contemporary humans).

Charles Darwin is the one who proposed the scientific argument for the theory of evolution by means of natural and sexual selection. According to this theory, successive generations of biological populations undergo changes in the inherited characteristics (genetic change in the gene pool). Over a large number of years, evolutionary processes give rise to diversity of life on Earth. Therefore, existing patterns of biodiversity have been shaped both by speciation and by extinction. Flora and fauna evolve and then sometimes die off, ceding their place on the planet to better-adapted species.

(h). Intellectual evolution

There is nothing like knowledge for its own sake. Knowledge and values are inextricably intertwined. Intellectual faculties are indispensable in the continued existence of the human species as they enable humans to address difficult challenges by envisioning the possible outcomes of specific actions.

The human capacity for the imaginative construct is apparent in the current state of knowledge. Scholarly science is an evolutionary practice. Key historical developments in erudite science include Darwin's theory of evolution; the discovery of genetics; the human genome; the germ theory of disease; the discovery of gas; the discovery of the chemical elements and the concept of atomic theory; Newton's theory of universal gravitation and classical mechanics; electro-magnetism; and Einstein's theories of special and general relativity. Other possible breakthroughs include sustainable habitats and possibly interstellar flight.

Since the dawn of humanity people have imagined the possibility of immortality in the context of an afterlife, heaven, hell, purgatory and karma. History is replete with out-of-body incidents, near-death experiences and other related reports and phenomena. It appears that, the current life-span is inadequate. Death takes its toll through accidents, suicide, disease and or old age. Prolonging health and or life, possibly indefinitely is something that has plagued humans since the dawn of humanity.

Right from prehistoric times, humans managed to develop techniques for making poisonous plants edible. Advances in curative and preventive medicine had their roots in the practice of Alchemy. Today food fortification is a strategy to reduce the incidence of nutrient deficiencies at the global level.

We go all-out desperately to postpone death for as long as possible. Major strides have been made in research on life extension in recent years. The average human life span has increased from less than 30 years in the medieval times to 80 years today. Scientific research indicates that living for 1000 years or more is a possibility. However, human beings may not be satisfied with this for they desire to extend life indefinitely.

People desire to live for countless years. Yet so far only a species of Planarian worms have managed to possess the capacity for indefinite longevity. Some scientists are looking for a 'death hormone' to make it possible for living systems to routinely repair themselves such that they do not succumb to disease or old age. Besides, latest scientific research has demonstrated that it is possible to create life from skin cells!

iii. Science is Random Phenomena

Uncertainty (randomness) is an integral component of nature. Reality is random (i.e. local and nonlocal). Reality is local (everyday experience) and non-local (evolution). The birth of the universe marked the beginning of science. Science incorporates everyday experience and adaptation. The primeval event and every other successive event(s) are instantaneous (faster than light in a vacuum) but the effects are observed gradually over time in the form of experience and adaptation.

Human beings as individuals have the capacity to tame the environment for their survival. However, today majority of the seven billion human beings on earth majority are unemployed thus dependant on the few who happen to be economically productive. Mainstream economic discussions of zero forced idleness since the 1970s suggest that attempts to reduce the level of joblessness below the natural rate of unemployment will fail, resulting only in less output and more inflation.

Unemployment is an anomaly. Unemployment is artificial and arises from restrictions on labour and money supply.

In the past, restrictions on labour took the form of primitive communalism, slavery, forced labour (poll, hut and breast tax) and colonialism. Today, it takes the form of migration laws and globalization (restrictions based on the post world war II matrix). The prevalent thinking is that unrestricted labor can depress wages in certain industries and create unemployment.

Money supply is susceptible to the shocks of business cycles (recessions and depressions). It is only in a single world currency regime that it is possible to adopt either an expansionary or contractionary monetary policy on a world scale. An expansionary policy increases the total supply of money in the economy while a contractionary policy shrinks it. An expansionary policy is necessary to combat unemployment in a recession while a contractionary policy is necessary to militate against inflation in order to avoid subsequent distortions and deterioration of asset values.

For the goal of science to be realized all without exception must participate, where possible. Fortunately, humans are naturally inclined to exertion. Even children are inseparable from child play.

Human beings are inclined to toil. Everyone spends some time in idleness now and then, but prolonged periods of idleness can have adverse physical and psychological effects on a person. The effects include obesity, loneliness, depression and cardiovascular disease.

Forced idleness occurs when humans beings are denied the opportunity to participate because of market failures such as economic fluctuations, unemployment, non-democratic forms of political systems, language diversity, gender structured power units (i.e. lineages, clans, ethnic groups, religions, races and civilizations), terrorism, war and of course early death.

Majority of the scientists across Africa, the Middle East, Asia, Australia, Asia and the Americas feel they are peripheral in the scientific process because of widespread blanket statements such as "Western knowledge", Western science", "Western forms of knowing", Western media", Western tradition" etc necessitating a rethink.

Scientists of all shades have been working to develop powerful tools and technologies to manage the fate of the universe.

Science is random phenomena. Science as an academic and practical activity is such that it pays generously when everyone is invited to play a part. Failure by any member to take part is costly. Therefore, knowledge must be made available and accessible to the wider public and the community of scholars for the insights to be put to use.

iv. Science is Partial

The poet Homer pioneered the idea that the struggles and triumphs of daily life follow a pattern. In scientific terms, natural, human and supernatural phenomenon obeys a universal law.

Science as an adaptive capacity is anchored on the indispensable altruism that the principle of the universe is to sustain life. The universe was meant to last by means of unending adaptation. Adaptation involves the interactions of physical, chemical, geological, and biological components of the physical world.

Human physical and mental activities (including academic science) impact on biodiversity and sustainability. Science as an adaptive capacity has a definitive goal to achieve immortality (connectedness). It is this quest that prompts scientists to ponder on the size of space, the possibility of traveling faster than light, the practicability of interstellar travel, the existence of God, the ultimate origins of humanity, the end of the world, and whether we are alone in the universe.

Human beings have changed the biosphere substantially in the quest to discover the unknown, reach the stars, reach other worlds and expand the human presence beyond, counter aging and achieve sustainable habitats.

Since the dawn of humanity death was (and remains) beyond human understanding and explanation. Nevertheless, humans' value life and that is why they pray, walk, run, eat, sleep, play, dream, read, search, think, create, share, conserve and imagine. To be sure, evidence abounds indicating that all living life (not only humans) are in search of the secret code of the universe – the life hormone.

Throughout history, humans' foremost desire is live endlessly. This truth is evident in science as an academic activity and as a practice: mystical and religious pursuits; art; literature; witches, medieval alchemists and medical practitioners; theoretical physicists, politicians; war merchants;

technological advancements; historians and their curiosity about the past; and many others.

The evolution of funerals is part of the effort to blur the line between dead and alive. Ancient humans attempted mummification and embalming with the hope of reviving their dead in the future. The search for signs of life or its key ingredients on Mars now or in the past is informed by the urge to preserve life and the fear of the unknown (death). We are preoccupied with clues to present and past Martian life for we hope to discover the deepest laws of the universe (nature).

Secular futurists envision technologies that may suspend death indefinitely. Such technologies include nanotechnology, artificial intelligence, robotics, aerospace research, engineered negligible senescence, and mind uploading (the transfer of the mind's information to a machine), among others.

Today humans are experimenting with guns that can Fire GPS-Guided Projectiles hoping to increase lethality while reducing collateral damage.

Non-Proliferation Treaty (NPT) aims to control the acquisition of the knowledge, capability and materials to build a nuclear bomb if it wants to. The International Atomic Energy Agency was established in 1957 to regulate the peaceful use of atomic energy and to ensure compliance to the Nuclear Non-Proliferation Treaty.

The world is gearing up for laws that penalize errant leaders who drag their countries into civil wars and acts of aggression against other states (including genocide and other crimes against humanity).

It will soon be standard procedure for all vehicles to mount the new innovative collision warning emergency braking system designed to help drivers avoid or minimize the rear-end collisions with stationary and moving vehicles to prevent serious accidents caused by negligence.

Witchcraft is an adaptive capacity. Witchcraft is misapplied futuristic technology. However, its association with ritual killings and secrecy made it to lose credibility. Human body parts have been used as ingredients for magical concoctions and charms. To obtain body parts for use in the occult, organ harvesters kill people. Today occult killing is murder, a capital offence. But witchcraft hasn't disappeared. Nevertheless, in places where it is deeply rooted, the regions remain largely underdeveloped as it tends to discourage investment and any form of personal development. Witchcraft is behind every misfortune, from infertility and poverty to failure in business, famine and earthquakes.

Sleep is an important adaptive capacity. Scientists have sought to understand why humans sleep for centuries in the attempt to suspend, halt or reverse the ageing and/or death process. During sleep humans experience partial or total unconsciousness. Consciousness has a lot to do with experience (sensation, emotions, thought and awakeness). Consciousness is the wakeful awareness of our own existence in an inextricably intertwined environment.

Remarkably, sleep and wakefulness are opposites and interdependent. Limited perception of environmental stimuli is one drawback associated with sleep. During sleep we partially or totally fail to respond to stimuli since our muscle activity, nervous system, senses and mental activity appears suspended. However, the benefits of sleep are well-known. Growth and repair occurs during sleep. Besides, after sleeping one feels relaxed, calm and energized. These are only two benefits but there are many more. Indeed, though sleep disorders and deprivation (sleeplessness) can be blessing in some circumstances, in most cases it is a nuisance or even a health setback. Death occurs when the failure to perceive the surrounding universe is total and permanent.

In the quest for unending life, humans have now touched and altered every part of the planet. Influences on biophysical systems as a result of human population growth with attendant resource consumption, habitat transformation and fragmentation, energy production and consumption, and climate change is evident. But so far, human beings only manage to attain a fleeting glimpse of immortality in art, scenic beauty, music, fantasy, science fiction, theology and in heightened emotion especially love.

Human behavior is impacted by its environment. Our very existence restricts our behavior. Self-preservation (more appropriately collective preservation) is behavior that ensures the continuity of life. For Darwin, avoidance of death is an evolutionary imperative (Darwin, 1859). According to Samuel Butler, self-preservation is the first law of nature. In reality, the preservation instinct is simply a demonstration of the foremost law of nature. The decree of the universe is to maintain life.

Mourning is the most evident expression of the human longing to evade passing away. Humans display profound grief - a general concern for the plight of others. Stories about grief stricken companion animals abound raising the idea that animals have a mental concept of death. Indeed, evidence indicate that wild animals such as giraffes, baboons, gorillas and elephants loiter around the body of a recently deceased close relative or loved one as they grieve those that have passed. Dolphins have been seen struggling to save a dead infant and to wail pitifully afterward. Nonhuman primates also grieve the loss of nonrelatives, including those who aren't genetically related them. Elephants, for example, extend this compassion to humans.

The universe is designed in such a manner that we have to serve others and the environment for us to sustain our species. For instance, in the household, local, regional and world economy we have to produce, distribute and trade in goods and services for the benefit of others to remain alive.

Religious entrepreneurs such as Jesus of Nazareth, the Buddha and Mohammed went further and dared to think beyond to imagine a life beyond this one. They imbibed piety, self-righteousness and preached the benefits of being of service to others as the key to this life and the life thereafter.

At the heart of their social thought was the idea of God (or gods) – one of the unresolved mysteries in science. 'Does God exist?' scientists often ponder. Indeed, the idea of God is a puzzle that has vexed humankind for centuries.

Science started as wonder, mystery, miracle or supernatural phenomenon. The idea of God (omnibenovelent, omnipotent, omnipresent, omniscience etc.) is synonymous with science. Albert Einstein is a religious reincarnate when he ponders on the possibility of travelling faster than the speed of light in a vacuum. Attaining such a feat will raise the possibility of experiencing the past, the present and the future simultaneously. Science is the quest to live forever. Science is an adaptive capacity.

The universe was meant to sustain life endlessly. Everything in this universe is connected. Science seeks to grasp the principle underlying human connectedness with the rest of the universe (or their environment).

For centuries, humans have used a common-sense approach to protecting and conserving natural resources such forests, water catchment areas. For instance, in the 1980s, the thinning ozone layer was one of the biggest environmental challenges facing humanity. In a considerable show of global cooperation the world ratified the Montreal Protocol and 25 years later, the ozone layer that protects the earth from the sun's most harmful ultraviolet rays is 98 percent phased-out. This collective fight against climate change has prevented millions skin cancer related deaths. The move has also eliminated ozone-depleting substances that are also potent greenhouse gases.

Apparently the natural world also obeys the law of the universe. For example, oceans curb the pace of climate change by absorbing carbon dioxide emissions from burning fossil fuels. That is why vast amounts of carbon are locked away in the depths of the Southern Ocean. This is simply another sign that the most important decree of the universe is to safeguard existence.

The world's roughly 7000 known languages are disappearing for the same reason. Language loss is the consequence of the gradual progression towards linguistic homogeneity as universal connectedness increasingly becomes a reality. The quest to sustain life is also what makes the sun to shine.

Death is the most dramatic episode that marks the end of our life experience. Death is a loss as it deprives us of experiences, milestones and all the things we value. Predictably bereavement induces pain and sorrow. Mourning is the human disappointment with death. Mourning take many forms including sobbing, crying, wailing, keeping vigil and/or a memorial service.

The universe is acceptably custom-made for the emergence and sustenance of life. If the universe was meant to endure how comes human beings are mortals? This is the problem of science.

The principle and practice of science revolves around life and death. Science helps us to understand and manipulate the world to extend our lifespans. Science (academic reasoning and experienced life) hopes to uncover the secret code of the universe and manipulate human destiny through continuous adaptation.

Scientists ask: what sparked life? Are we alone in the universe? Is the inner Earth hollow with openings at its poles? What happens to us when we die? What is the size and shape of space? Is it possible to turn back time? Is it possible to travel forward in time? Is time travel possible? Plainly put: Is it possible to live forever?

Life is a journey, not a destination. The problem of science is how to permit life forms to overcome exceptional odds to stay alive enduringly. Human beings aspire to live and live on.

CONCLUSION

Science is an adaptive capacity. Science encompasses academic scholarship tinkered with experience. Science is also evolutionary; arbitrary; and purposeful.

CHAPTER THREE

CONCLUSION

This book has demonstrated that science is not simply supernatural phenomenon, mythology and tradition. Science can also not be satisfactorily explained as the search for good judgment. Further science is not merely the body of knowledge of the natural world and/or the process by which that knowledge is acquired. Rather science is defined by the purpose. Science is about what use systematic knowledge is put to. Science is a change process. Science is an adaptive capacity.

How knowledge is acquired (the scientific method) is only important to validate and replicate the findings.

Different scientific disciplines typically use different methods and approaches to investigate the natural world. Even then, branches of learning in the natural sciences and the human sciences have similarities. Being adaptive capacities, they share a common goal.

The worth of a scientific engagement is not measured by the many observational or experimental tests it passes. Science is useful insofar as it enables life forms to appreciate the intricate laws of nature necessary to improve and sustain life permanently.

Science as an adaptive capacity incorporates scholarship[3] and experience[4]. Science is prompted by experience. The most unwelcome experience is that of death. From early on, living life would rather bereavement was not part of their experience. Science is the theory and practice of harnessing the natural world to stay alive.

A cursory examination of life-forms (and non-life forms) may mislead one into concluding that they are working at cross-purposes. For example, it is common to hypothesize that life-forms battle for power, resources and mates. Actually, the universe is connected by a shared purpose. The design of science is to sustain life endlessly.

---

[3] Knowledge acquired by study
[4] Knowledge acquired by doing something, through participation or observation

BIBLIOGRAPHY

Adams Fred& Laughlin Gregory (2000). *The Five Ages of the Universe: Inside the Physics of Eternity*. Simon & Schuster Australia.

Allen, Thomas George (1974). "The Book of the Dead or Going Forth by Day: Ideas of the Ancient Egyptians Concerning the Hereafter as Expressed in Their Own Terms". *SAOC* vol. 37; University of Chicago Press.

Aristotle (2007) [350 BC]. *Metaphysics*. NuVision Publications, LLC.

Aristotle (350 BCE). *Physics*. Hardie RP, Gaye RK (Trans). Available at: http://classics.mit.edu/Aristotle/physics.html (accessed 26 December 2013).

Cave Stephen (2012). *Immortality: The Quest to Live Forever and How It Drives Civilization*. Crown Publishing Group.

Chomsky Naom (1965). *Aspects of the Theory of Syntax*. MIT Press.

Chomsky Naom (1975). *Reflections of Language*. New York: Pantheon Books.

Chomsky Naom (2000). *New Horizons in the Study of Language and Mind*. Cambridge: Cambridge University Press.

Clement Catherine (1983). *The Lives and Legends of Jacques Lacan*. New York: Columbia University Press.

Collingwood R.G. (1946). *The Idea of History*. Oxford: Oxford University Press.

Darwin Charles (1859). *On the Origin of Species by Means of Natural Selection, or the Preservation of Favoured Races in the Struggle for Life*. London: John Murray.

Grant Edward (1996). *The Foundations of Modern Science in the Middle Ages: Their Religious, Institutional, and Intellectual Contexts*. Cambridge: Cambridge University Press.

Ellis J (2000). "There's a place for the theory of everything". *Nature* 403, 241-242.

Feyerabend Paul (1975). *Against Method: Outline of an anarchistic theory of knowledge*. London: Verso.

Foucault Michel (1978). *The History of Sexuality, Volume 1: An Introduction*. Random House.

Freud Sigmund (2001). *The Complete Psychological Works*. Strachey J (ed). Vintage Paperbacks.

Fried Morton (1972). *The Notion of the Tribe*. Cummings Pub. Co.

Harrison Edward (2003). *Masks of the Universe: Changing Ideas on the Nature of the Cosmos*. Cambridge University Press.

Hawking Stephen W. & Leonard Mlodinow (2010). *The Grand Design*. New York: Bantam Books.

Hawking Stephen W. (1988). *A Brief History of Time*. New York: Bantam Books.

Hawking Stephen W. (2001). *The Universe in a Nutshell*. New York: Bantam Spectra.

Hayek Friedrich (1944). *The Road to Serfdom*. Routledge.

Heidegger Martin (1978). *Being and Time*. Wiley-Blackwell.

Herodotus (1987) [484 BC - 425 BCE]. *The Histories*. Trans. David Grene. University of Chicago Press.

Jackson Roy (2001). *Plato: A Beginner's Guide*. London: Hoder & Stroughton.

Kant Immanuel (1848) [1781]. *Critique of Pure Reason*. London: William Pickering.

Khamala Geoffreyson (2009). "Gender Dimension of Ethnic Conflicts in Kenya: The Case of Bukusu and Sabaot Communities". MA Thesis, Kenyatta University, Kenya.

Khamala Geoffreyson (2014). *The Perfect Theory: A Complete Unified Description of the Universe*. Tajiriba Foundation.

Kovacs Maureen Gallery (1989). *Epic of Gilgamesh*. Trans. Maureen Gallery Kovacs. Stanford University Press.

Kuhn Thomas (1962). *The Structure of Scientific Revolutions*. Chicago: The University of Chicago Press.

Lacan Jacques (1977). *Écrits: A Selection*. Trans. Alan Sheridan. London: Tavistock Publications.

Marshall M (2010). Knowing the mind of God: Seven theories of everything. *NewScientist*. Available at: http://www.newscientist.com/article/dn18612-knowing-the-mind-of-god-seven-theories-of-everything.html?

Mbiti John (1969). *African Religion and Philosophy*. Heinemann.

Nagy Gregory (2010). *Homer: the Preclassic*. Berkeley: University of California Press.

Piaget Jean (1937/1955). *The Construction of Reality in the Child.* London: Routledge & Kegan Paul.

Plato (380 BC). *The Republic.* Trans. Benjamin Jowett. BompaCrazy.com.

Plato [1928]. *Phaedo.* Trans. Benjamin Jowett. Forgotten Books.

Popper Karl (1936). *The Poverty of Historicism.* Routledge & Kegan Paul.

Popper Karl (1945). *The Open Society and Its Enemies.* Routledge.

Popper Karl R. (2002). [1959]. *The Logic of Scientific Discovery.* New York, NY: Routledge Classics.

Rawls John (1971). *A Theory of Justice.* Cambridge, Mass.: Belknap Press of Harvard University Press.

Rawls John (2001). *Justice as Fairness: A Restatement.* Cambridge, Mass.: Harvard University Press.

Schopenhauer Arthur (1958). *The World as Will and Representation.* Falcon's Wing Press.

Selengut Charles (2003). *Sacred Fury: Understanding Religious Violence.* Walnut Creek, CA: Altamira.

Service Rogers Elman (1975). *Origins of the State and Civilization: The Process of Cultural Evolution.* Norton.

Skinner BF (1957). *Verbal Behavior.* Acton, MA: Copley Publishing Group.

Thucydides [1946]. *History of the Peloponnesian War.* Trans. Sir Richard Winn Livingstone & Richard Crawley. Forgotten Books.

Weinberg S (1993). *Dreams of a Final Theory: The Search for the Fundamental Laws of Nature.* London: Hutchinson Radius.

Wikipedia (2014). "Science".
http://en.wikipedia.org/wiki/Science. Retrieved Saturday, November 15, 2014.

www.ingramcontent.com/pod-product-compliance
Lightning Source LLC
Chambersburg PA
CBHW051737170526
45167CB00002B/967